An Introduction to Manufacturing and Mathematical Limitations by James Jesse Mills

Copyright © 2024 by James Jesse Mills.

Library of Congress Control Number: 2024915373
ISBN: Softcover   979-8-3694-2644-9
             eBook   979-8-3694-2627-2

All rights reserved. No part of this book may be reproduced or transmitted in any form or by any means, electronic or mechanical, including photocopying, recording, or by any information storage and retrieval system, without permission in writing from the copyright owner.

Any people depicted in stock imagery provided by Getty Images are models, and such images are being used for illustrative purposes only.
Certain stock imagery © Getty Images.

Print information available on the last page.

Rev. date: 04/15/2025

To order additional copies of this book, contact:
Xlibris
844-714-8691
www.Xlibris.com
Orders@Xlibris.com
861477

Foreword

That which is crooked cannot be made straight, and that which is wanting cannot be numbered.
Ecclesiastes 1:15.

Consider the work of God: for who can make that straight, which he hath made crooked?
Ecclesiastes 7:13

Many of the following are applications of Ecclesiastes (1:15, 7:13), written by Solomon. If the word "straight" is thought of as "the original" or "the perfect," then these verses in Ecclesiastes may be applied to modern technology.

Preface

Much of this book is in simple or simplified terminology, and the mathematics is not rigorous. I hope that you will find this an interesting read.

Table of contents

Some mechanical limitations
Some electrical limitations
Some mathematical limitations
Some more mathematical limitations

This book may contain errors. Please do not use this book for critical applications such as airplanes, boats, medicines, hospital equipment, etc. Please use standard references for critical applications.

If you are disturbed by "infinite" and "zero" words, replace them with unattainable numbers.

The number six has been overused to reduce problems with numerology and the fear of numbers. If this is disturbing, feel free to use different numbers. Some of the mathematics may need to be different.

According to the second law of thermodynamics and other laws relating to machinery and electronics, it is impossible to manufacture perfect things. However, it is possible to manufacture things that are close to perfection.

Some mechanical limitations of cars, boats, and airplanes (jet and propeller)

Rust and tarnish.
The maximum power of motors cannot be infinite.
The maximum speed of vehicles cannot be the speed of light.
The maximum speed at which vehicles can safely turn cannot be the speed of light.
It is not possible to manufacture perfect wheels. Nobody manufactures products with a tolerance of plus or minus zero. For example, it is impossible to manufacture a wheel with a radius of 8.000000…… inches, with infinite zeros after the decimal point.
It is not possible to align wheels perfectly. Nobody manufactures wheel alignment equipment that can align wheels with a plus or minus zero tolerance.
The divergence of laser beams cannot be zero.
The angle between the front wheels and the angle of the rear wheels on modern cars is adjustable. The angle of both front wheels cannot be exactly the same when turning or moving straight. For example, the angle of both front wheels cannot be exactly 13.000000…… degrees, with an infinite number of zeros after the decimal point.
The angle of both rear wheels cannot be exactly 0.000000…… degrees, with infinite zeros after the decimal point.
The distance between the two front wheels cannot be exactly the same as the distance between the two rear wheels.

When moving straight, the distance between the two left wheels cannot be exactly the same as that between the two right wheels.
When moving straight, the angle between the front and side wheels cannot be exactly 90 degrees.
When moving straight, the angle between the rear and side wheels cannot be exactly 90 degrees.
The braking resistance on each wheel is not exactly the same. This is one of the reasons why tires need to be rotated.
Four-wheel drive vehicles do not have the exact same torque on each wheel. Especially between the front wheels and the rear wheels, and when turning. This is another reason why tires need to be rotated.
The two driven wheels on two-wheel drives with differentials may not have the exact same amount of torque when turning. This is another reason to rotate tires periodically. When turning, the outer wheel revolves more than the inner wheel. Differentials can allow one wheel to coast while turning. There are also other methods for solving this problem. In conventional rear-wheel drive vehicles, the differential connects the drive shaft to the left and right halves of the rear axle.
Four-wheel drives with differentials may have the same limitations that are described above.
It is not possible to manufacture tires that do not wear out.
It is not possible to manufacture tires that cannot skid or slide.
It is not possible to manufacture perfectly round spheres.
It is not possible to manufacture perfectly straight rods or shafts. A perfectly straight rod or shaft can be considered an infinite number of perfect circles that are all exactly the same size or an infinite

number of perfect squares that are all exactly the same size.
Manufacturing bumpers that cannot be dented, bent, or cracked is impossible.
The friction of bearings cannot be zero.
Conversion losses apply to non-powered converters, which may be in powered equipment.
Conversion loss cannot be zero when fuel is converted into mechanical or electrical energy.
The conversion loss cannot be zero when mechanical energy is converted into electrical energy.
The conversion loss cannot be zero when electrical energy is converted into mechanical energy.
The conversion loss cannot be zero when wind energy is converted into electrical energy.
The conversion loss cannot be zero when water flow energy is converted into electrical energy.
The conversion loss cannot be zero when electrical energy is converted into heat energy.
The conversion loss cannot be zero when heat energy is converted into electrical energy.
The conversion loss cannot be zero when electrical energy is converted into light energy.
The conversion loss cannot be zero when light energy is converted into electrical energy.
The conversion loss of hydraulics cannot be zero.
The conversion loss when converting to different RPMs cannot be zero.
The conversion loss of gear drives cannot be zero.
The conversion loss of belt drives cannot be zero.
The conversion loss of chain drives cannot be zero.
The conversion loss of shaft drives cannot be zero.
There is a conversion loss when kinetic energy is put onto the shaft and when kinetic energy is removed.

An Introduction to Manufacturing and Mathematical Limitations

It is not possible to manufacture stick-shift transmissions that shift gears perfectly smoothly. For example, making a perfectly smooth shift from first to second gear is impossible.

Oils cannot reduce friction to zero.

Oil filters cannot remove all the impurities from oils.

The error of speedometers (the number of digits after the decimal point on the output display cannot be infinite), odometers (the number of digits after the decimal point on the output display cannot be infinite), tachometers, fuel gauges, temperature gauges, pressure gauges, vacuum gauges, altimeters, and other types of measuring instruments cannot be ± 0.

The regulation of cruise controls cannot be plus or minus zero. For example, it is not possible to maintain a constant speed of 45.000000...... miles per hour with an infinite number of zeros after the decimal point.

The delay before airbags open cannot be zero.

It is impossible for spark plugs to spark at exactly the same time every time, especially at high RPMs. Spark plugs should spark at approximately 3 degrees past the maximum compression point as the piston moves away from the end of the cylinder. For example, it is impossible to manufacture an engine where the spark plugs spark every time at 2.95000000...... degrees past the maximum compression point (measured at the crankshaft), with infinite zeros after the number 5.

It is not possible for intake and exhaust valves to open and close at exactly the same times (measured at the crankshaft) every time, especially at high RPMs.

Carburetors do not mix fuel into the air perfectly evenly.

Fuel injectors do not mix fuel into the air perfectly evenly.
The leakage of piston rings cannot be zero.
Friction between piston rings and cylinder walls cannot be zero.
Friction between pistons and cylinder walls cannot be zero.
The leverage of piston rods and the crankshaft is not the same at all rotation points.
It is not possible to make perfect flywheels. It is impossible to remove the ripples from the output of piston engines absolutely and completely.
The conversion loss of alternators cannot be zero.
The output of alternators is a series of positive and adjacent sine wave halves, which look like rectified alternating current. Most of the ripples are filtered out with a voltage regulator. It is impossible to completely filter out the ripples from the output of alternators and AC adapters, which internally have the same waveform.
Suppose a very sensitive, very low-frequency radio is connected through a circuit that separates the ripples and prevents overdriving the radio to the output of the best AC adapters or power supplies; in that case, it will still be able to detect a 50/60 Hertz ripple.
The following is a simplified explanation of switching AC adapters and power supplies:
Switching AC adapters and power supplies converts 50/60 Hertz AC into direct current. A transistor switches the direct current on and off to form square waves with a much higher frequency than 50/60 Hertz. The square waves are then converted into alternating current with a transformer, rectified, and filtered. Voltage is regulated by automatically adjusting the length of

the on/off times of the switching transistor. The higher frequency makes using smaller transformers, smaller filtering capacitors, and smaller inductors possible. Switching AC adapters and power supplies can be compact and lightweight. It is not possible to completely filter the ripples from the output of switching AC adapters and power supplies.

The output of electric motors is not perfectly smooth.

The output of electric motors that are powered by 50/60 Hertz AC has a 50/60 Hertz ripple. Brushes connect the current to the electromagnets that are on the revolving armature.

The output of electric motors powered by direct current has ripples when the commutator(brushes) switches the electromagnets from north to south and vice versa.

The output of brushless electric motors powered by direct or rectified alternating current is not perfectly smooth.

The ripples from the output of piston engines and electric motors can be reduced by reducing the RPM with gear drives, chain drives, or belt drives. If the ripples from the output of piston engines and electric motors are almost completely filtered out by reducing the RPM with gear drives, chain drives, or belt drives, the output RPM would be almost zero.

Mufflers do not reduce noise to zero. Note that modern cars can be very quiet. Ordinary and common mufflers cannot make the exhaust flow from piston engines perfectly smooth.

The volume of toxic gases at the output of catalytic converters cannot be zero.

Automotive shocks do not reduce shocks to zero. Making rides over speed bumps and potholes

perfectly smooth is impossible.
The maximum weight that trucks can carry cannot be infinite.
The maximum weight that boats can carry cannot be infinite.
The maximum weight that airplanes can carry cannot be infinite.
The minimum weight of airplanes and helicopters cannot be zero.
Resistance to airflow(drag) on the surface of airplanes and cars moving through still air cannot be zero.
Resistance to airflow beneath the main rotor blades of helicopters cannot be zero.
Helicopter tails' resistance to airflow from the tail rotor blades cannot be zero.
Resistance to water flow(drag) on the hull of boats moving through still water cannot be zero.
It is not possible to make perfect propellers. Perfect propellers would go through water or air like bolts go through nuts.
It is not possible to manufacture perfect rudders. Perfect rudders would enable boats to make sharp turns like cars.
It is not possible to build perfect airplane wings. When the speed of airplanes is too slow, their wings do not have enough lift, and when the speed of airplanes is too fast, their wings may have too much lift.
The left and right wings cannot be perfectly symmetrical.
It is not possible to manufacture perfect airplane tails. Perfect airplane tails would enable airplanes to make sharp turns like cars.
The melting temperature of metals cannot be infinite.

The maximum weight that steel beams can hold cannot be infinite.
The maximum weight that nuts and bolts can hold cannot be infinite.
Headlamps do not make roads without streetlights or roads in heavy fog at night as visible as roads in sunny daylight.
It is not possible to make windows that cannot break or crack.
It is not possible to make perfectly clear windows. Perfectly clear windows would be invisible at all angles and close distances.
In stormy weather, windshield wipers do not wipe all the rain off windshields.
Light energy absorbed in mirrors cannot be zero.
The blind spots of ordinary or common rearview mirrors cannot be zero.
Video cameras may have blind spots.

Some electrical limitations

Switches cannot be turned on and off an infinite number of times.

The maximum number of times digital flip-flops can toggle cannot be infinite. Transistors and other circuit components eventually wear out for various reasons. Digital flip-flops are one of the main building blocks of computers. They can store a logic one or a logic zero. Digital flip-flops can be wired to toggle (change from a logic zero to a logic one, or vice versa) with each square wave at the clock input, so two square waves at the clock input will produce one square wave at the output. A two-mega-Hertz square wave at the clock input will produce a one-mega-Hertz square wave at the output, and a two-giga-Hertz square wave at the clock input will produce a one-giga-Hertz square wave at the output. Flip-flops can be combined with logic gates to divide by any number.

The maximum frequency at which digital flip-flops can toggle cannot be infinite.

The maximum frequency at which digital frequency counters can count is not infinite.

Data(a zero or a 1) must be present at the input of a flip-flop for a minimum amount of time before the digital clock signal loads the data into the flip-flop.

Data(a zero or a 1) must remain at the input of a flip-flop for a minimum amount of time after the digital clock signal loads the data into the flip-flop. This is one of the reasons computers cannot do anything instantaneously.

The delay of tri-state buffers when electrically connecting the output of digital devices to data buses cannot be zero.

The delay of tri-state buffers when electrically disconnecting the output of digital devices from data buses cannot be zero. Tri-state buffers can allow multiple digital devices to share the same data bus. The output of a tri-state buffer can be the same as the digital device connected to its input, or the output of the tri-state buffer can be in electrically disconnected mode(similar to an open switch).

The maximum clock frequency of microprocessors cannot be infinite.

The maximum number of tasks microprocessors can perform simultaneously cannot be infinite.

Maximum data transfer rates cannot be infinite.

The delay between the inputs and outputs of digital flip-flops cannot be zero.

The delay between the inputs and outputs of logic gates cannot be zero.

The delay before computers can do anything cannot be zero. It takes time to form square waves.

Address bus widths cannot be infinite.

Data bus widths cannot be infinite.

The length of time of reads and writes in solid-state drives, hard drives, and other types of digital storage cannot be zero.

The capacity of solid-state drives, hard drives, and other types of digital storage cannot be infinite.

The resistance of wires cannot be zero.

The delay of signals through wires cannot be zero.

The delay of computer data buses cannot be zero.

The distortion of wires cannot be zero. If there are previous points of sinusoidal-type signals at the end of wires because of the delay of wires, there will be a small amount of distortion. Varying charge densities form sinusoidal points in wires. A higher-than-normal density (zero volts) of electrons produces a negative sinusoidal point, and a lower-than-normal density of electrons will produce a positive sinusoidal point.

Avoid using excessively long speaker cables, microphone cables, radio frequency coaxial cables, microwave waveguides, computer data cables, printed circuit board traces, etc.

Noise is the unwanted addition to signals traveling through wires, audio cables, coaxial cables, waveguides, computer data buses, printed circuit board traces, switches, and other components.

The noise of wires in circuits cannot be zero. The unwanted electrical influence of nearby circuit components or nearby wires may cause noise in a wire.

The maximum current that wires can carry cannot be infinite.

Diodes and transistors are made by adding chemicals to silicon so that the silicon atoms have more than the normal amount of electrons (N material) or the silicon atoms have fewer than the normal amount of electrons (P material). If N material is joined to P material to form a diode, a negative voltage is applied to the P material, and a positive voltage is applied to the N material, the diode will have a high resistance. But if a positive voltage is applied to the P material and a negative voltage to the N material, the diode will have a low resistance.

NPN bipolar transistors can be formed by sandwiching a P material between two N materials. A wire is connected to the middle P material, called the base, and wires are connected to the N materials. If a voltage causes a current to flow through one of the diodes in the forward direction, then a much larger current can flow through both diodes, even though one is in the reverse direction. A small current through the base can control a much larger current through both diodes. Current does not flow through the diode in the reverse direction when there is no base current. This applies to PNP transistors also. Bipolar transistors are manufactured so that one of the diodes works better in the forward direction and the other diode works better in the reverse direction. In NPN transistors, the connection to the N material of the diode that works better in the forward direction is called the emitter, and the connection to the N material of the diode that works better in the reverse direction is called the collector. The connection to the middle semiconductor material common to both diodes is called the base. This also applies to PNP transistors.

Field-effect transistors (FETs) can be formed by joining a P or N material, called a gate, to a narrow silicon channel of the opposite material. If a voltage is applied between the gate and the end of the channel that is called the source so that the junction between the gate and the channel has similarities to a diode that is biased in the reverse direction, then an electrostatic field with similarities to the electrostatic fields in capacitors is formed that reduces the electrical width of the already narrow silicon channel so that the resistance of the channel is higher. If the voltage is high enough, the current through the channel can be cut off. A small amount of input power at the gate can control a much larger output power through the channel. MOSFETs are formed by putting a metal oxide layer between the gate and the channel so that less direct current flows through the gate into the channel. Field effect transistors are manufactured so that one channel end works better as the source. The other end is called the drain.

The following is a simplified explanation of a vacuum tube used for amplification:

An electrical conductor called a cathode is heated to a high temperature by a heater filament similar to filaments in light bulbs. The cathode is connected to a negative voltage. Another electrical conductor, the plate, is connected to a positive voltage. Electrons can easily leave the hot cathode, and current can flow from the negative cathode to the positive plate. A metal grid between the cathode and the plate is connected to a voltage that is negative with respect to the voltage at the cathode. The negative grid reduces the current flow from the cathode to the plate. If the grid voltage is high enough, the current flow from the negative cathode to the positive plate can be cut off. A small amount of grid current can control a much larger amount of current flowing from the cathode to the plate. A larger, amplified sinusoidal signal can appear at the plate if a small sinusoidal signal is applied to the grid.

Electrons going through bipolar transistors produce more noise than electrons going through wires, since the output current goes through two diode junctions in bipolar transistors. Also, electrons going through FETs and vacuum tubes produce more noise than electrons going through wires. Generally, bipolar transistors (NPNs and PNPs) have less distortion than FETs and tubes, which are the correct type for the same application. FETs and tubes that cannot handle as much current or voltage may have less distortion. However, FETs and tubes generally have less noise than bipolar transistors, which are the correct type for the same

application. Bipolar transistors that cannot handle as much current or voltage may have less noise.
The forward resistance of diodes cannot be zero.
The forward voltage of diodes cannot be zero.
Typical forward voltages are from 0.2 volts to 1.2 volts.
The maximum forward current of diodes cannot be infinite.
The reverse resistance of diodes cannot be infinite.
The maximum reverse voltage of diodes cannot be infinite.
The reverse current of diodes cannot be zero.
Mixers are used to down-convert signals from antennas to a lower frequency, called the intermediate frequency. For example, a 1.455 mega-Hertz signal from an oscillator (sine wave generator) in the receiver can be mixed with a one mega-Hertz radio signal (amplitude modulated) from the antenna to produce a 455 kilo-Hertz intermediate frequency. A 110.7 mega-Hertz oscillator (sine wave generator) in the receiver can be mixed with a 100 mega-Hertz radio signal (frequency modulated) from the antenna to produce a 10.7 mega-Hertz intermediate frequency. Different radio stations have different frequencies, but the intermediate frequency in receivers is still the same.
The delay before diodes can turn on in the forward direction cannot be zero. When diodes turn on, the current increases at a higher and higher rate (not a constant rate) until the maximum current is reached. The increasing(nonconstant) rate is useful for mixing microwave signals. When diodes are used for switch mixing (mixing lower frequency radio signals), the delay before the oscillator signal reaches the forward voltage to turn on the diodes produces a small amount of distortion.

The maximum signal frequency that diodes can rectify cannot be infinite.
The delay of electron beams through vacuum tubes cannot be zero.
The delay of electromagnetic radio signals in the atmosphere cannot be zero.
The delay between the input and the output of transistors, vacuum tubes, and other types of amplification devices cannot be zero.
The delay before fuses interrupt circuits cannot be zero.
The delay before ground fault circuit interrupters can interrupt circuits cannot be zero.
The sensitivity of ground fault circuit interrupters cannot be infinitesimal (infinitely small).
The distortion of transistors, vacuum tubes, and other types of amplification devices cannot be zero.
The noise of transistors, vacuum tubes, and other types of amplification devices cannot be zero.
The input and output capacitance of transistors, vacuum tubes, and other types of amplification devices cannot be zero.
The input and output inductance of transistors, vacuum tubes, and other types of amplification devices cannot be zero.
The maximum output power of transistors, vacuum tubes, and other types of amplification devices cannot be infinite.
The capacitance between the wires in audio cables (speaker cables, microphone cables, etc.) cannot be zero.
When one type of energy is converted into another, the conversion loss cannot be zero.
The conversion loss cannot be zero when audio energy is converted into electrical energy.
The conversion loss cannot be zero when electrical

energy is converted into audio energy.
The conversion loss cannot be zero when electrical energy is converted into electromagnetic energy.
The conversion loss cannot be zero when electromagnetic energy is converted into electrical energy.
Conversion loss when a voltage or current is converted into a different voltage or current cannot be zero. There are too many conversion losses to make a complete list.
The resistance and inductance of capacitors cannot be zero.
The maximum capacitance of capacitors cannot be infinite.
The maximum voltage that can be put across capacitors cannot be infinite.
The resistance and inductance of voltage-controlled variable capacitors cannot be zero.
The distortion of sinusoidal signals that go through electrolytic capacitors cannot be zero. Electrolytic capacitors do not charge and discharge exactly the same as capacitors with two plates and a nonconductive material between the two plates.
The resistance and capacitance of inductors cannot be zero.
The maximum inductance of inductors cannot be infinite.
The maximum amount of current that can go through inductors cannot be infinite.
The delay before the magnetic cores of inductors and transformers can change their direction from north to south (and vice versa) cannot be zero.
There is an upper-frequency limit where the magnetic core fails to keep up with the signal in the coil around it, and a lower-frequency limit where the number of turns to produce commonly used

inductance values near the upper-frequency limit would be too great.
The distortion of signals that go through magnetic core inductors cannot be zero.
The distortion of signals that go through air-core inductors cannot be zero. See distortion of wires.
Electrical signals that go through inductors or capacitors may be distorted. Electrical signals converted into electromagnetic or electrostatic fields and then converted back into electrical signals may not be exactly the same. High-performance audio amplifiers do not have many inductors or capacitors in the signal's path.
The flux loss of inductors and transformers cannot be zero.
The hysteresis loss in the magnetic cores of inductors and transformers cannot be zero.
The eddy current in the magnetic cores of inductors and transformers cannot be zero.
The distortion of magnetic core transformers cannot be zero.
The distortion of air core transformers cannot be zero. See distortion of wires.
Air-core transformers attenuate low-frequency signals more than high-frequency signals.
Phase distortion of radio frequency signals with information on them that go through air core or magnetic core transformers cannot be zero. A radio signal without information is just a sine wave.
The capacitance between the turns of air core and magnetic core inductors cannot be zero. When the frequency is above the usable frequency of the inductor, too much of the signal goes through the capacitance between the turns.
The maximum input frequency of transformers cannot be infinite.

Power lost in the magnetic cores of inductors and transformers cannot be zero.

The conversion loss of transformers cannot be zero. For example, if the input power of a transformer is 10 watts, the output power will be less than 10 watts.

The maximum input and output voltage of transformers cannot be infinite.

The maximum input and output current of transformers cannot be infinite.

The maximum input and output power of transformers cannot be infinite.

The maximum power that resistors can dissipate cannot be infinite.

The error of voltmeters, ammeters, ohmmeters, inductance meters, capacitance meters, power meters, frequency counters, SWR(standing wave ratio) meters, and other measuring instruments cannot be plus or minus zero.

Over a long period, magnets will lose their magnetism.

The input impedance of quarter-wave antennas is low because the input signal bounces off(reflected) the end of the antenna and neutralizes the signal at the input a half wavelength later. Suppose the input signal is removed from the input of an antenna. In that case, the signal already in the antenna bounces back and forth between the ends of the antenna(resonance), producing a sinusoidal-type signal that decreases to zero volts.

The input of a quarter-wavelength antenna is an electrical summing point. There is a slight and negligible amount of distortion because the reflected signal is half of a wavelength delayed ($\frac{1}{4} + \frac{1}{4} = \frac{2}{4} = \frac{1}{2}$). This problem can be reduced by decreasing the reflected signal, which will increase the input impedance of the antenna. The impedance of the antenna should be increased until it is the same as the impedance of the coaxial cable connected to it, usually 50 ohms or 75 ohms.

The ringing (resonance) of antennas cannot be zero. Resonance tends to remove the information from sinusoidal-type signals and make them more like perfect sine waves. This problem can also be reduced by decreasing the reflected signal, which will increase the input impedance of the antenna.

The distortion of antennas cannot be zero. See distortion of wires.

The ringing (resonance) of coaxial cables cannot be zero.

The distortion of coaxial cables cannot be zero.

The ringing of waveguides cannot be zero.

The distortion of waveguides cannot be zero.

The following two paragraphs explain resonance:

If a sine wave generator with a high output impedance is connected to a capacitor and an inductor that are in parallel. At a particular frequency, called the resonant frequency, when the reactance (alternating current resistance of capacitors and inductors) of the parallel capacitor and inductor are at the same magnitude, the parallel capacitor and inductor will resonate and have a high impedance.

If a sine wave generator with a low output impedance is connected to a capacitor and an inductor that are in series. At a particular frequency, called the resonant frequency, when the reactance (alternating current resistance of capacitors and inductors) of the series capacitor and inductor are at the same magnitude, the series capacitor and inductor will resonate and have a low impedance.

Distortion of radio signals with information on them that are at or near the resonant frequency of the inductors they go through cannot be zero. The capacitance between the turns causes resonance. The transitions from a logic 0 to a logic 1, or vice versa, in digital signals are not perfectly perpendicular (as seen on an oscilloscope) because it takes at least a small amount of time to transition from a digital 0 to a digital 1, or vice versa. This adds ambiguity to digital signals and is one reason why building perfect digital modulators and demodulators is impossible.

It is not possible to form perfect square waves.

It is not possible to form perfect sawtooth waves.

It is not possible to form perfect sine waves.

It is impossible for digital signals taken from radio signals to be exactly the same as the original digital signals. The length of the ones and/or the length of the zeros will not be exactly the same.

The distortion and noise of amplitude modulators cannot be zero.

The distortion and noise of amplitude demodulators cannot be zero.

The distortion and noise of frequency modulators cannot be zero.

The distortion and noise of frequency demodulators cannot be zero.

It is not possible to build perfect analog-to-digital converters.

It is not possible to build perfect digital-to-analog converters.

It is not possible to build perfect video cameras.

Digital audio: The number of bits per sample cannot be infinite. In between the numbers the bits represent is audio information that cannot be represented.

The number of samples per second cannot be infinite. Audio information is missing between the samples.

Digital video: The number of bits per pixel cannot be infinite. In between the numbers the bits represent is picture information that cannot be represented.

The number of pixels per frame(picture) cannot be infinite. In between the pixels is missing picture information.

The number of frames(pictures) per second cannot be infinite. In between the frames is missing video information.

The wow and flutter of CD, record, and tape players cannot be zero.

The crossover distortion of push-pull amplifiers cannot be zero. One transistor amplifies the positive half of the sine wave or sinusoidal-type wave, and another amplifies the negative half of the sine or sinusoidal-type wave. There is a small amount of distortion as the signal goes from the positive half of the sine or sinusoidal-type wave to the negative half of the sine or sinusoidal-type wave and vice versa.

Single-transistor or single-vacuum tube amplifiers have a different type of distortion. Even if a perfect sine wave is at the input of a single transistor or single vacuum tube amplifier, the positive half of the sine wave at the output of the transistor or vacuum tube amplifier will still not be perfectly symmetrical with the negative half of the sine wave at the output of the transistor or vacuum tube amplifier. The amplification or gain of the transistor or vacuum tube changes as the input signal changes.

Push-pull amplifiers may use matching

complementary transistors. It is not possible to obtain exactly matching complementary PNP and NPN bipolar transistors or exactly matching complementary P-channel and N-channel FETs that have the exact same amount of gain(amplification), the same maximum voltage, the same maximum current, the same maximum output power, etc. Distortion when using vacuum tubes to simulate PNP transistors or P-channel MOSFET transistors in push-pull vacuum tube audio amplifiers cannot be zero. Electron current goes in the opposite direction through PNP transistors or P-channel MOSFETs to amplify half of the sine or sinusoidal-type waves. It is impossible to manufacture vacuum tubes, so the electron current goes in the opposite direction, from the plate to the cathode. It is possible to make vacuum tubes simulate PNP transistors or P-channel MOSFETs, but not exactly.

The distortion and noise of oscillators cannot be zero. It is not possible to generate perfect sine waves. It is possible to reduce the distortion of sine waves to a tiny amount, but not to absolute zero.

The frequency drift of oscillators cannot be zero. The frequency drift of quartz crystal oscillators cannot be zero. Quartz crystals can be used to build very stable, low-drift, but not zero-drift, oscillators. For example, the output of a two-megahertz quartz crystal oscillator can be divided two million times to produce square waves with very accurate one-second 0(s) and very accurate one-second 1(s). However, quartz clocks and watches are still not as accurate as the atomic clock, which has an error of less than one second in two billion years.

The phase noise of oscillators cannot be zero. Phase noise is a common problem of phase-locked loop oscillators used in digital radios. Frequency instability in oscillators can cause phase noise. For example, a sine wave is phase-shifted forward at the transition point of a frequency increase, and at the transition point of a frequency decrease, a sine wave is phase-shifted backward.

The phase noise of radios and televisions cannot be zero.

Filtering can cause phase distortion. For example, a low-pass filter will phase-shift high-frequency components more than low-frequency components, and a high-pass filter will phase-shift low-frequency components more than high-frequency ones.

The delay before voltage-controlled oscillators change to different frequencies cannot be zero. This is another reason why building perfect digital modulators is impossible.

Impedance mismatch between RF amplifiers and impedance matching networks cannot be zero, with infinite zeros after the decimal point.

Impedance mismatch between impedance-matching networks and loads cannot be zero.

Impedance mismatch between the transmission lines and impedance-matching networks cannot be zero.

Impedance mismatch between impedance-matching networks and antennas cannot be zero.

Impedance mismatch between RF amplifiers in integrated circuits cannot be zero, especially if external impedance matching is not used.

Amplifiers impedance-matched to the source and load for maximum output power may have more distortion.

The distortion of speakers cannot be zero. Audio signals go through two transformations when they are converted into sound. The first transformation is from electrical to electromagnetic energy (speaker coil). The second transformation is from electromagnetic energy to kinetic energy (the moving of the speaker cone). It is not possible for these two transformations to be exact.

The distortion of microphones cannot be zero.

All audio signals have at least a small amount of distortion and noise.

It is impossible to transform electrical energy into electromagnetic energy in antennas exactly.

The transformation of electromagnetic energy to electrical energy in antennas cannot be exact.

The sound from a radio cannot be exactly the same as that heard at the radio station.

The picture from a television cannot be exactly the same as that seen on the television station.

Electric keyboards cannot sound exactly the same as real pianos. Combining various audio oscillators to produce perfect piano sounds is impossible. High-performance electric keyboards use the recordings of real pianos.

The input impedance of voltmeters cannot be infinite.

The internal impedance of ammeters (impedance between the input and output) cannot be zero. The impedance that the ammeter adds to the tested circuit cannot be zero.

The internal impedance of standing wave ratio (SWR) meters cannot be zero.

The internal impedance of in-line RF power meters cannot be zero.

The input impedance of amplifiers cannot be infinite.

The output impedance of amplifiers cannot be zero.
The input impedance of amplifiers cannot be zero.
Some circuits work better when the amplifier has a low input impedance.

The output impedance of amplifiers cannot be infinite. Some circuits work better when the amplifier has a high output impedance.

The input and output capacitance of amplifiers cannot be zero. The input and output capacitance of amplifiers is one of the factors that determine the maximum frequency that amplifiers can amplify.

The delay of negative feedback in audio amplifiers cannot be zero. Negative feedback is commonly used in audio amplifiers to reduce distortion. If the delay of the negative feedback is zero, it would be possible to build nearly perfect audio frequency amplifiers.

When signals go through passive filters, energy loss cannot be zero. Passive filters are filters made with inductors, capacitors, and resistors.

Distortion and noise are a fraction of the output of filters made with inductors, capacitors, and resistors. To reduce distortion and noise to zero, the signal would need to be filtered until it is zero (a fraction of zero is zero).

To reduce distortion and noise on a radio signal to zero, it must be filtered until it is a pure sine wave. The information that is on it would be completely lost.

Filters cannot remove distortion and noise from audio signals absolutely and completely without also losing the audio information completely. This concept is used when analog (not digital) and digital audio devices are manufactured. Distortion and noise are not absolutely and completely filtered out because there would be nothing left to listen to. This is one of the reasons why very low-distortion audio electronics have humongous bass. The treble parts of the audio are reduced or lost during filtering. Most of the distortion and noise is in the treble parts of the audio.

It is impossible to absolutely and completely remove the ripples from the digital steps of digital audio without losing the audio information completely. If a very sensitive, very low-frequency radio is connected through a circuit that separates the ripples and prevents overdriving the radio to the output of the best CD player, MP3 player, cell phone, computer, digital signal processor, or other source of digital audio, it will still be able to detect the ripples from the digital steps.

It is impossible to completely filter out distortion and noise absolutely and completely from the output of radio frequency (RF) mixers without losing the analog or digital information.

The regulation of voltage regulators cannot be ± 0.

The maximum amount of energy stored in batteries cannot be infinite.

The maximum output voltage of batteries cannot be infinite.

The maximum output current of batteries cannot be infinite.

The internal impedance of batteries cannot be zero.

The maximum number of times rechargeable batteries can be recharged cannot be infinite.

The mechanical and electrical sections are easy.

Nothing that is mechanical or electrical is perfect.

Some mathematical limitations

The following mathematical concepts may have limitations, but they are still useful and important concepts, and should not be neglected in mathematical studies. Some of the following are not limitations, but are for educational purposes.

The following examples are approximations because they are irrational numbers. Irrational numbers are numbers that have an infinite number of nonzero digits after the decimal point. The last digit is rounded to the nearest digit. Errors are negligible. Some of the irrational numbers in this article are described as exact for simplification. The actual numbers that are used are not infinitely long because of computer limitations and the limitations of measuring equipment.

π (π is the ratio between the diameter and the circumference of circles. π is also the rotational distance of half circles, and 2π is the rotational distance of full circles. π is $3.141592......$ radians, and π is equal to 180 degrees. Radians may work better than degrees when the variable need to be similar to real numbers. Real numbers are the set of rational and irrational numbers from $-\infty$ to $+\infty$.

$e(e = 2.718281......)$ The derivative of e^x is e^x, and the integral of e^x is $e^x + C$.

Not all functions are the same when differentiated. Not all functions are the same when integrated.

Euler's constant

$\frac{2}{3}$

$cosine\ 56°$ The derivative of $cosine\ t$ is $-sine\ t$, and the derivative of $-sine\ t$ is $-cosine\ t$. After the

second differentiation, the function is the same, but the sign is reversed(negative).
A similar derivative is the derivative of $\cosh t$(hyperbola), which is $\sinh t$, and the derivative of $\sinh t$ is $\cosh t$. The function is the same after the second differentiation, and the sign is still the same (positive).

$sine$ 56° The derivative of $sine\ t$ is $cosine\ t$, and the derivative of $cosine\ t$ is $-sine\ t$. After the second differentiation, the function is the same, but the sign is reversed(negative).
A similar derivative is the derivative of $\sinh t$(hyperbolic sine), which is $cosinh\ t$, and the derivative of $cosinh\ t$ is $\sinh t$. The function is the same after the second differentiation, and the sign is still the same(positive).

$tangent$ 56°
$\log 56$
$\ln 56$
Planck's constant
mass of a muon
mass of a neutron
mass of a proton
mass of an electron
charge of an electron
radius of electrons
frequency of photons
permittivity in a vacuum constant
Avogadro's number
magnetic flux quantum constant
Compton wavelength constant
permeability constant
Dirac's constant
fine structure constant
Bohr magneton constant

Bohr radius constant
nuclear magneton constant
2π
π^2
e^2

It is not possible to calculate the length of curved lines or the circumference of circles exactly. The area between the curved line and the straight lines (the measuring method) represents a small amount of error, regardless of how short the straight lines are. A computer can be used to divide the curved line into many segments for a more accurate approximation. This is one of the methods to calculate the number π. There may be easier methods to calculate the number π, but the results are still an irrational number. The number π can be used to easily calculate the circumference, area, and volume of circles, spheres, and cylinders.
Curved lines that are formed with many short-straight lines may not be exact.
Straight lines that are formed with many short-curved lines may not be exact.
Circles that are formed with many short-straight lines may not be exact.
Spheres that are formed with many small flat surfaces may not be exact.
It may not be possible to exactly calculate the area beneath some curved lines.
It may not be possible to exactly calculate the volume beneath some curved surfaces.
The following integrations and differentiations are examples of the fundamental theorem of calculus and Stokes theorem involving circles and spheres:
$A = \int C i r d r$ where in symbolic integration, dr means to differentiate the function r to obtain the

derivative, which is $1 (dr = nr^{n-1} = 1r^{1-1} = r^0 = 1)$
A is the equation of the area of circles
Cir is the equation of the circumference of circles
$A = \int 2\pi r\, dr$ use the formula $\dfrac{r^{n+1}}{n+1} + C$, where n is an integer, and multiply by 2π
$A = \dfrac{2\pi r^{1+1}}{1+1} = \dfrac{2\pi r^2}{2}$
$A = \pi r^2$, and
$Cir = dA$
Cir is the equation of the circumference of circles
A is the equation of the area of circles
$Cir = d\pi r^2$ use the formula nr^{n-1}, where n is an integer, and multiply by π
$Cir = 2\pi r^{2-1}$
$Cir = 2\pi r$, also
$V = \int S\, dr$ where in symbolic integration, dr means to differentiate the function r to obtain the derivative, which is $1 (dr = nr^{n-1} = 1r^{1-1} = r^0 = 1)$
V is the equation of the volume of spheres
S is the equation of the surface area of spheres
$V = \int 4\pi r^2\, dr$
$V = \dfrac{4\pi r^{2+1}}{2+1}$
$V = \dfrac{4\pi r^3}{3}$, and
$S = dV$
S is the equation of the surface area of spheres
V is the equation of the volume of spheres
$S = d\dfrac{4\pi r^3}{3}$
$S = \dfrac{(4)(3)\pi r^{3-1}}{3} = \dfrac{(4)(3)\pi r^2}{3}$
$S = 4\pi r^2$

The following is a simplified explanation of *cosine* and *sine*:
The *cosine* of an angle is the ratio of x and the

radius of circles at that particular angle, and the *sine* of an angle is the ratio of *y* and the radius of circles at that particular angle. Following are some examples:

cosine 0(0 radians is 0 degrees) $= \frac{x}{r} = \frac{1}{1} = 1$ (The radius is 1 for simplification.)

sine $0 = \frac{y}{r} = \frac{0}{1} = 0$ (The radius is 1 for simplification.)

When the center of the circle is at the origin($x = 0, y = 0$), then 0 radians is the positive *x* axis.

cosine $\frac{\pi}{4}(\frac{\pi}{4}$ radians is 45 degrees) $= \frac{x}{r} = \frac{\frac{1}{\sqrt{2}}}{1} = \frac{1}{\sqrt{2}}$

sine $\frac{\pi}{4} = \frac{y}{r} = \frac{\frac{1}{\sqrt{2}}}{1} = \frac{1}{\sqrt{2}}$

cosine $\frac{\pi}{2}(\frac{\pi}{2}$ radians is 90 degrees) $= \frac{x}{r} = \frac{0}{1} = 0$

sine $\frac{\pi}{2} = \frac{y}{r} = \frac{1}{1} = 1$ When the center of the circle is at the origin($x = 0, y = 0$), then $\frac{\pi}{2}$ radians is the positive *y* axis.

The equations for circles that are centered at the origin are ($x^2 + y^2 = r^2$ where *r* is the radius), and (*rcosine t* + *risine t*, where *r* is the radius, and *t* is the angle). When *t* goes a distance of 2π, a full circle is drawn. The function *rcosine t* replaces the *x* axis (*x* = *rcosine t*), and *i* indicates that the function *risine t* replaces the imaginary or *y* axis (*y* = *rsine t*).

For circles with a radius other than 1, simply multiply the *cosine* value and the *sine* value with the different radius value. For example, the equation for circles that are centered at the origin with a radius of 2 is 2*cosine t* + 2*isine t*, where *t* is the angle, the function 2*cosine t* replaces the *x* axis(*x* = 2*cosine t*), and *i* indicates that the function 2*isine t* replaces the imaginary or *y* axis(*y* = 2*sine t*).

The equation for spheres is: $x^2 + y^2 + z^2 = r^2$. Spheres can also be drawn by trigonometric functions by replacing the x axis with $(sine\ u)(cosine\ t)$, replacing the y axis with $(sine\ u)(sine\ t)$, and by replacing the z axis with $cosine\ u$. In other words, $x = (sine\ u)cosine\ t)$, $y = (sine\ u)(sine\ t)$, and $z = cosine\ u$. The independent variable u increments or decrements between 0 and π in steps, and the independent variable t goes a distance of 2π at each step. This draws a circle with the correct radius at each step. These functions draw a sphere that is centered at the origin with a radius of one.

It may not be possible to describe curved lines, circles, spheres, or curved surfaces exactly with numbers. Some points on curved lines, circles, spheres, or curved surfaces would need to be described with irrational numbers. For example, when x, and/or y is equal to $\sqrt{2}$, $\sqrt{3}$, $\sqrt{5}$, or $\sqrt{6}$.

The equation for lines is $y = ax + b$, where a is the slope of the line and b is where the line crosses the y axis. For example, if $y = 2x + 2$, then the slope of the line is 2 and the line crosses the y axis at 2.

It may not be possible to describe straight lines exactly with numbers. Some points on straight lines would need to be described with irrational numbers. For example, when x and/or y is equal to $\sqrt{2}$, $\sqrt{3}$, $\sqrt{5}$, or $\sqrt{6}$.

Limits may be approximations. For example, when x is infinity, the limit of $\frac{(4x+800)}{x}$ is not exactly equal to 4.

The following is the infinite series of $0.999999......$ and $1.999999......$, and the infinite product of $1.999999......$:

$1 \cong$ ($\cong$ means approximately equal to)

An Introduction to Manufacturing and Mathematical Limitations

$0.999999\ldots\ldots = \sum_{n=1}^{\infty} \frac{1}{2^n} = \frac{1}{2^1} + \frac{1}{2^2} + \frac{1}{2^3} + \frac{1}{2^4} + \frac{1}{2^5} + \frac{1}{2^6} + \cdots$

$2 \cong 1.999999\ldots\ldots = \sum_{n=1}^{\infty} \frac{2}{2^n} = \frac{2}{2^1} + \frac{2}{2^2} + \frac{2}{2^3} + \frac{2}{2^4} + \frac{2}{2^5} + \frac{2}{2^6} + \cdots$

$2 \cong 1.999999\ldots\ldots = \prod_{n=1}^{\infty} \frac{(1+\frac{1}{n})^2}{1+\frac{2}{n}} =$

$(\frac{(1+\frac{1}{1})^2}{1+\frac{2}{1}})(\frac{(1+\frac{1}{2})^2}{1+\frac{2}{2}})(\frac{(1+\frac{1}{3})^2}{1+\frac{2}{3}})(\frac{(1+\frac{1}{4})^2}{1+\frac{2}{4}})(\frac{(1+\frac{1}{5})^2}{1+\frac{2}{5}})(\frac{(1+\frac{1}{6})^2}{1+\frac{2}{6}})\cdots$

Integers and rational numbers in infinite series forms or infinite product forms may have errors that are the size of the errors of irrational numbers. Rational numbers do not have an infinite number of nonzero digits after the decimal point.
Integer points and rational points of functions that are in infinite series or infinite product forms may have errors that are the size of the errors of irrational numbers.
Infinite series and infinite product version of the Gamma function (Γ). The Gamma function (Γ) is the factorial of the number that is one less. For example, the Gamma function of 5($\Gamma(5)$) is 4! = (4)(3)(2)(1) = 24, and the Gamma function of 6($\Gamma(6)$) is 5! = (5)(4)(3)(2)(1) = 120.
Many infinite series and infinite products have errors that are the size of the errors of irrational numbers(negligible) and may be considered as exact.
Multivariate functions that are converted into infinite series may not be exact.
Newton iterations for calculating the square root of integers.
Numerical calculations are calculations that replace variables with numbers.

A derivative is a function that is the slope of the original function. For example, the derivative of $\frac{x^3}{3}$ is x^2 (use the formula nx^{n-1}, where n is an integer, and divide by 3): $\frac{d\ x^3}{dx\ 3} = \frac{nx^{n-1}}{3} = \frac{3x^{3-1}}{3} = \frac{3x^2}{3} = x^2$). The slope of $\frac{x^3}{3}$ at $x = 1$ is 1 ($x^2 = 1^2 = 1$). Another example is, the slope of $\frac{x^3}{3}$ at $x = 2$ is 4 ($x^2 = 2^2 = 4$).

An integral is the opposite of a derivative. For example, the integral of x^2 is $\int x^2 dx = \frac{x^3}{3} + C$ (use the formula $\frac{x^{n+1}}{n+1} + C$, where n is an integer). In symbolic integration, dx means to differentiate the function x to obtain the derivative, which is 1 ($dx = nx^{n-1} = 1x^{1-1} = x^0 = 1$). $\int x^2 dx = \frac{x^{n+1}}{n+1} + C = \frac{x^{2+1}}{2+1} + C = \frac{x^3}{3} + C$)

Another example is the integral of y^2 which is $\int y^2 dy = \frac{y^3}{3} + C$ (Use the formula $\frac{y^{n+1}}{n+1} + C$, where n is an integer.) In symbolic integration, dy means to differentiate the function y to obtain the derivative, which is 1 ($dy = ny^{n-1} = 1y^{1-1} = y^0 = 1$).

$\frac{x^3}{3}$ is an example of a differential 0-form. $x^2 dx$ is an example of a differential 1-form and a covariant tensor. Another example of a differential 1-form and a contravariant tensor is $f(x, y, z)dx + g(x, y, z)dy + h(x, y, z)dz$, where f, g, and h are functions with the variables x, y, and z. $f(x, y, z)$ is pronounced f of x, y, and z. Similarly, for $g(x, y, z)$, and $h(x, y, z)$. When the original function is differentiated with respect to x, the differential form is $f(x, y, z)dx$. When the original function is differentiated with respect to y, the differential form is $g(x, y, z)dy$, and when the original function is differentiated with respect to z,

the differential form is $h(x, y, z)dz$.
An example of a differential 2-form and a contravariant tensor is $f(x, y, z)dydz + g(x, y, z)dzdx + h(x, y, z)dxdy$, where f, g, and h are functions with the variables x, y, and z. When the original function is differentiated, the differential forms are either $f(x, y, z)dydz$, $g(x, y, z)dzdx$, or $h(x, y, z)dxdy$.

The following are examples of f, g, and h for differential 2-forms obtained with an averaging method. Other averaging methods may also be used. It may be necessary to replace the variables with numbers and numerically square root:

$$f(x, y, z) = \sqrt{\frac{(\frac{d}{dy}(original\ function))^2 + (\frac{d}{dz}(original\ function))^2}{2}}$$

f may not be exact for some functions.

$$g(x, y, z) = \sqrt{\frac{(\frac{d}{dz}(original\ function))^2 + (\frac{d}{dx}(original\ function))^2}{2}}$$

g may not be exact for some functions.

$$h(x, y, z) = \sqrt{\frac{(\frac{d}{dx}(original\ function))^2 + (\frac{d}{dy}(original\ function))^2}{2}}$$

h may not be exact for some functions.
A differential form is not exact if it cannot be equal to the derivative of another differential form, and a differential form is exact if it is equal to the derivative of another differential form. Exact differential forms are also closed. Closed differential forms are equal to zero when differentiated. Some closed differential forms are not exact. Differential forms that are closed and not exact may be analyzed with cohomological methods.
Differentiation of multivariate functions with respect

to multiple variables may not be exact.
Integration of multivariate functions with respect to multiple variables may not be exact.
Integration of multivariate differential forms with respect to multiple variables may not be exact.
Differential forms may not be exact when dx, dy, and dz are not of infinitesimal lengths. In numerical calculations, dx means a small difference in x, dy means a small difference in y, and dz means a small difference in z.
Some tangent bundles or vector bundles, which are commonly used in differential forms, may be approximations.
Between tangent bundles or vector bundles may be missing information.
Some Jacobian matrices may be approximations.
The area between a straight tangent line (df) and a short segment of the curved function (Δf) may not be zero.
The following definite integrals are examples of the fundamental theorem of calculus, and Stokes theorem:

$\int_1^2 x^2 \, dx$ exactly calculates the area between x^2 and the horizontal axis from $x = 1$ to $x = 2$:

To calculate $\int_1^2 x^2 \, dx$, first integrate x^2:

$$\int x^2 \, dx = \frac{x^3}{3}$$

Then applying the integral of x^2:

$\int_1^2 x^2 \, dx$
$= F(2) - F(1)$ where F is the integrated x^2
$= \frac{2^3}{3} - \frac{1^3}{3}$
$= \frac{8}{3} - \frac{1}{3}$
$= \frac{7}{3}$

$= 2\frac{1}{3}$ square units.

$\int_1^2 x^3 \, dx$ exactly calculates the area between x^3 and the horizontal axis from $x = 1$ to $x = 2$:

To calculate $\int_1^2 x^3 \, dx$, first integrate x^3:

$$\int x^3 \, dx = \frac{x^4}{4}$$

Then applying the integral of x^3:

$\int_1^2 x^3 \, dx$
$= F(2) - F(1)$ where F is the integrated x^3
$= \frac{2^4}{4} - \frac{1^4}{4}$
$= \frac{16}{4} - \frac{1}{4}$
$= \frac{15}{4}$
$= 3\frac{3}{4}$ square units.

Following is an example of an integration with respect to two different variables at the same time:

$$\iint f(x, y, z) \, dy \, dz$$

Following is an example of a definite integration with respect to two different variables at the same time:

$\int_a^b \int_c^d f(x, y, z) \, dy \, dz$ calculates the volume beneath $f(x, y, z)$ from $y = a$ to $y = b$ and from $z = c$ to $z = d$.

Some Riemann integrations may be approximations. Riemann integrations are numerical calculations of definite integrals. Riemann integrations are the numerical calculation of the area beneath lines using small squares, and the numerical calculation of the volume beneath surfaces using small cubes. In numerical integration, dx means a small difference in x, dy means a small difference in y, and dz means a small difference in z.

Calculation of the area beneath curved lines using many small triangles may not be exact.
When integrating, it may be necessary to omit undefined points, or apply other types of restrictions before or after integration. For example, if the function, or integrated function is $\frac{5x}{\sqrt{x-4}}$, then x needs to be greater than 4. If the function is integrated with respect to only one variable, the other variables are held constant during integration. Not all symbolic expressions can be symbolically integrated. For example, x^2 and x^3 can be symbolically integrated ($\frac{x^3}{3} + C, \frac{x^4}{4} + C$), but $x^{2\frac{1}{3}}$, $x^{2\frac{1}{2}}$, and $x^{2\frac{2}{3}}$, cannot be symbolically integrated.
Not all symbolic expressions can be symbolically differentiated or numerically differentiated. For example, it may not be possible to symbolically, or numerically differentiate fractals.
When differentiating, it may be necessary to omit undefined points, or apply other types of restrictions before or after differentiation. For example, if the function or differentiated function is $\frac{5x}{\sqrt{x-4}}$, then x needs to be greater than 4. If the function is differentiated with respect to only one variable, the other variables are held constant during differentiation.
Symbolic expressions that cannot be symbolically integrated can be numerically integrated. Symbolic expressions that are numerically integrated may be approximations. The area between the integrated function and the tangent lines, and the distance between the points may represent a small amount of error. In numerical integration, dx means a small difference in x, dy means a small difference in y,

and dz means a small difference in z.
Symbolic expressions that cannot be symbolically differentiated can be numerically differentiated. Symbolic expressions that are numerically differentiated may be approximations. The area between the differentiated function and the tangent lines, and the distance between the points may represent a small amount of error.
Numerical methods may not be exact.

Some more mathematical limitations

The following mathematical concepts may have limitations, but they are still useful and important concepts, and should not be neglected in mathematical studies.

Application of Simpson's rule for definite integrals may not be exact.

Integration of functions produce variable constants. For example, the integral of x is $\frac{x^2}{2} + C_1$. If x is integrated again, there are an infinite number of possible results, for example: $\int \frac{x^2}{2} + C_1 = \frac{x^3}{6} + C_1 x? + C_2?$ The constant C_1 in $\frac{x^2}{2} + C_1$ determines how high or low $\frac{x^2}{2}$ is relative to the horizontal axis. For example, if C_1 is 2, then the function would be 2 units higher with respect to the horizontal axis, if C_1 is -2, then the function is 2 units lower with respect to the horizontal axis. C_1 is omitted or assigned the value of zero in some of the integrations for simplification.

$\iiint$ usually means to integrate with respect to three different variables once each. $\iiint$ usually does not mean to integrate three times with respect to only one variable.

When integrating or differentiating with respect to only one variable, the other variables are held constant. Following are some examples:

$\int yx dx = y \frac{x^2}{2} + C$ (the $d\_$ notation indicates the variable to integrate with respect to).

$\int xy dy = x \frac{y^2}{2} + C$

$\frac{d}{dx}\left(y\frac{x^2}{2} + C\right) = yx + 0$ (The derivative of constants is zero.)$= yx$, (d means to differentiate the following function and $\frac{d}{dx}$ means to differentiate the following function with respect to x).

$\frac{d}{dx}\left(y\frac{x^2}{2} + y^2\right) = yx + 0 = yx$

$\frac{d}{dy}\left(x\frac{y^2}{2} + C\right) = xy + 0$ (The derivative of constants is zero.)$= xy$, (d means to differentiate the following function and $\frac{d}{dy}$ means to differentiate the following function with respect to y.)

$\frac{d}{dy}\left(x\frac{y^2}{2} + x^2\right) = xy + 0 = xy$

The following example is the differentiation of the function $(x^2 + y^2)$ with respect to two different variables. Note the absence of " $= r^2$".

$df(x,y) = d(x^2 + y^2)$
$= 2xdx + 2ydy$

When differentiating $x^2 + y^2$ with respect to x, the differential form is $2xdx$ ($\frac{d}{dx}y^2 = 0$), and when differentiating $x^2 + y^2$ with respect to y, the differential form is $2ydy$ ($\frac{d}{dy}x^2 = 0$).

When differentiating $x^2 + y^2$ twice, the derivative is zero. The derivative of many differential forms equal zero, or approximately equal zero. When the derivative of a differential form is zero, the differential form is called a closed differential form. Not all closed differential forms are exact, but all exact differential forms are closed. Differentiating twice for differential forms with respect to one variable while holding the other variables constant twice may not work.

$2xdx + 2ydy$ is an example of a contravariant

tensor.

$\begin{matrix} 2xdx \\ 2ydy \end{matrix}$ is an example of a covariant tensor.

The following example is the differentiation of the function $(x^2 + y^2 + z^2)$ with respect to three different variables. Note the absence of "$= r^2$".

$df(x, y, z) = d(x^2 + y^2 + z^2)$
$= 2xdx + 2ydy + 2zdz$

When differentiating $x^2 + y^2 + z^2$ with respect to x, the differential form is $2xdx$ ($\frac{d}{dx}y^2 = 0$ and $\frac{d}{dx}z^2 = 0$).

When differentiating $x^2 + y^2 + z^2$ with respect to y, the differential form is $2ydy$ ($\frac{d}{dy}x^2 = 0$ and $\frac{d}{dy}z^2 = 0$),

and when differentiating $x^2 + y^2 + z^2$ with respect to z, the differential form is $2zdz$ ($\frac{d}{dz}x^2 = 0$ and $\frac{d}{dz}y^2 = 0$).

When differentiating $x^2 + y^2 + z^2$ twice, the derivative is zero. The derivative of many differential forms equal zero, or approximately equal zero. When the derivative of a differential form is zero, the differential form is called a closed differential form. Not all closed differential forms are exact, but all exact differential forms are closed. Differentiating twice for differential forms with respect to one variable while holding the other variables constant twice may not work.

$2xdx + 2ydy + 2zdz$ is an example of a contravariant tensor.

$\begin{matrix} 2xdx \\ 2ydy \\ 2zdz \end{matrix}$ is an example of a covariant tensor.

The equation for the determinant of a four-term square matrix $\begin{pmatrix} a & b \\ c & d \end{pmatrix}$ is: $determinant = ad - bc$.

The inversion of a four-term square matrix $\begin{pmatrix} a & b \\ c & d \end{pmatrix}$

is: $\frac{1}{determinant} \begin{pmatrix} d & -b \\ -c & a \end{pmatrix}$

Some basic matrix equations are:

$Ay = B$ and

$y = \frac{1}{A}B$ where A is an original matrix, y is the solution, B is the output, and $\frac{1}{A}$ is the inverted matrix

For example, if A is $\begin{pmatrix} 6 & 4 \\ 8 & 6 \end{pmatrix}$, and the desired output is $\begin{matrix} 2 \\ 4 \end{matrix}$, then the solution is:

$y = \frac{1}{A}B$

$y = \frac{1}{(6)(6)-(4)(8)} \begin{pmatrix} 6 & -4 \\ -8 & 6 \end{pmatrix} \begin{matrix} 2 \\ 4 \end{matrix}$

$y = \frac{1}{36-32} \begin{pmatrix} (6)(2) & (-4)(4) \\ (-8)(2) & (6)(4) \end{pmatrix}$

$y = \frac{1}{4} \begin{pmatrix} 12 & -16 \\ -16 & 24 \end{pmatrix}$

$y = \begin{matrix} \left(\frac{1}{4}\right)(12) + \left(\frac{1}{4}\right)(-16) \\ \left(\frac{1}{4}\right)(-16) + \left(\frac{1}{4}\right)(24) \end{matrix}$

$y = \begin{matrix} 3 - 4 \\ -4 + 6 \end{matrix}$

$y = \begin{matrix} -1 \\ 2 \end{matrix}$

Following is the proof that $\begin{matrix} -1 \\ 2 \end{matrix}$ is the solution for the matrix $\begin{pmatrix} 6 & 4 \\ 8 & 6 \end{pmatrix}$ when the desired output is $\begin{matrix} 2 \\ 4 \end{matrix}$:

$Ay = B$

$Ay = \begin{pmatrix} 6 & 4 \\ 8 & 6 \end{pmatrix} \begin{matrix} -1 \\ 2 \end{matrix}$

$Ay = \begin{matrix} (6)(-1) + (4)(2) \\ (8)(-1) + (6)(2) \end{matrix}$

$Ay = \begin{matrix} -6 + 8 \\ -8 + 12 \end{matrix}$

$$Ay = \frac{2}{4}$$

It is not possible to obtain the determinant of matrices that are not square.

When the determinant of a matrix is not equal to zero, the matrix has a unique solution. The matrix can be inverted, and the unique solution is the inverted matrix multiplied with the result or output matrix. When the determinant of a matrix is equal to zero, the matrix has an infinite number of solutions, and does not have a unique solution, because at least two rows of the matrix are identical, or they differ only by a multiplicative factor (linearly dependent).

Two rows of a matrix are linearly dependent if they are identical, or if one row is a multiplication or fraction of the other row.

If a matrix has linearly dependent rows, the matrix has an infinite number of solutions, and does not have a unique solution.

Some matrices cannot be exactly inverted.

It is not possible to obtain solutions, or exact solutions for all matrices.

Row reduced matrices are not exactly the same as the original matrices.

The matrix $\begin{pmatrix} 3 & 6 \\ 6 & 12 \end{pmatrix}$ has a determinant of zero:

$determinant = (3)(12) - (6)(6) = 36 - 36 = 0$

A matrix with a determinant of zero has a volume size of zero.

The diagonal terms of a matrix can be modified with an eigenvalue so that the determinant is equal to zero. For example, when the eigenvalue 2 is added to the diagonal terms of the matrix $\begin{pmatrix} 1 & 6 \\ 6 & 10 \end{pmatrix}$, the determinant is equal to zero:

An Introduction to Manufacturing and Mathematical Limitations

$$\begin{pmatrix} 1+2 & 6 \\ 6 & 10+2 \end{pmatrix} = \begin{pmatrix} 3 & 6 \\ 6 & 12 \end{pmatrix}$$

It is not possible to obtain eigenvalues for matrices that are not square.

It may not be possible to obtain exact eigenvalues for large matrices.

Some matrices cannot be decomposed.

Some matrices cannot be exactly decomposed.

The following is a matrix and it's solution for unaltered circles:

$\begin{matrix} 1 & 0 \\ 0 & 1 \end{matrix} \; \begin{matrix} x \\ y \end{matrix}$ where x and y are the coordinate values from the equation $x^2 + y^2 = r^2$

The following is a matrix and it's solution for an ellipse that is twice as wide as it is high:

$\begin{matrix} 2 & 0 \\ 0 & 1 \end{matrix} \; \begin{matrix} x \\ y \end{matrix}$ where x and y are the coordinate values from the equation $x^2 + y^2 = r^2$

Following is a matrix and it's solution for unaltered spheres:

$\begin{matrix} 1 & 0 & 0 \\ 0 & 1 & 0 \\ 0 & 0 & 1 \end{matrix} \; \begin{matrix} x \\ y \\ z \end{matrix}$ where x, y, and z are the coordinate values from the equation $x^2 + y^2 + z^2 = r^2$

Following is a matrix and it's solution for a spheroid that is twice as wide as it is high:

$\begin{matrix} 2 & 0 & 0 \\ 0 & 2 & 0 \\ 0 & 0 & 1 \end{matrix} \; \begin{matrix} x \\ y \\ z \end{matrix}$ where x, y, and z are the coordinate values from the equation $x^2 + y^2 + z^2 = r^2$

x^2 in two dimensions is a curved line, but x^2 as a vector is a straight line.

x^3 in two dimensions is a curved line, but x^3 as a vector is a straight line.

Curved two-or-three dimensional functions that are converted into vectors may not be exactly the same.

Vectors that are converted into curved two-or three-dimensional functions may not be exactly the same. Curved two-or three-dimensional functions that are converted into vectors for Lagrangian multiplication may not be exactly the same.

In vector mathematics, other functions may replace the x, y, and z axis. This is not considered a limitation. For example, instead of the x axis, $2x$ may be used, instead of the y axis, $3x$ may be used, ($y = 3x$), and instead of the z axis, $2x^2$ may be used, ($z = 2x^2$). Then, when the independent variable x is 2, the coordinates of the vector are 4,6, and 8.

Vectors do not need to replace the x, y, and z axis. For some examples, two points of a vector can be:
The origin ($x = 0, y = 0$) and the point ($x = 8, y = 8$).
The origin ($x = 0, y = 0$) and the point ($x = 10, y = 9$).
The point ($x = 4, y = 2$) and the point ($x = 6, y = 8$).
The point ($x = 8, y = 6$) and the point ($x = 2, y = 4$).
The point ($x = 4, y = 2, z = 4$) and the point ($x = 6, y = 8, z = 6$).
The point ($x = 8, y = 6, z = 5$) and the point ($x = 2, y = 4, z = 8$).
Vector space matrices that are modified with eigenvalues are not exactly the same as the original matrices.
Vector space mathematics involving Fourier series transformations may be approximations.
Vectors can also represent the variables of a function. For example, if the variables of a function

are $\begin{smallmatrix} x \\ y \\ z \end{smallmatrix}$, then $\begin{smallmatrix} 4 \\ 2 \\ 4 \end{smallmatrix}$ means $x = 4$, $y = 2$, and $z = 4$.

Squares that are formed with curved functions may not be exact.

Cubes that are formed with curved functions may not be exact.

Rectangles that are formed with curved functions may not be exact.

Rectangular boxes that are formed with curved functions may not be exact.

The maximum number of vectors and terms(numbers) in tensors (vector mathematics) cannot be infinite. Between the vectors and terms(numbers) of tensors (vector mathematics) may, (but not necessarily) be missing information.

A contravariant vector is the same as a covariant vector if they describe the same mathematical object, except the contravariant vector describes the mathematical object differently than the covariant vector. If a tensor contains a contravariant vector and a covariant vector that describe the same mathematical object, there will be a cancellation.

Christoffel symbols involves averaging methods.

An undefined point is a point where the numerator is not zero and the denominator is zero. Nonzero numbers that are divided by zero are undefined. For example, if the function is $\frac{4x}{(x-2)}$, then the undefined point is 2. Points adjacent to undefined points are still too large for computers. For example, infinity minus one, infinity minus two, and infinity minus three, are too large for computers.

Going over the undefined points of integrated functions and calculating the distance or area may not be exact. For example, if the integrated function is $\frac{x}{x-5}$, then going over from 4 to 6 would not be exact, because the numerator would be less than 5 on one side of the undefined point, and the numerator would be greater than 5 on the other side of the undefined point. This would make the two adjacent areas of the undefined point not equal. The adjacent areas on both sides of undefined points are not always equal. Cauchy principal values make the two adjacent areas equal to each other. Functions that are modified by Cauchy principal values are not exactly the same as the original functions.

Exponent integral (Ei) involves a principal value. The exponent integral is $\int \frac{e^x}{x} dx$ (where in symbolic integration, dx means to differentiate the function x), and when x is negative: $\int \frac{e^{-x}}{-x} dx = \int \frac{1}{(e^x)(-x)} dx = \int \frac{1}{-xe^x} dx$. Ei is undefined at $x = 0$. Note that $e^0 = 1 \cong e^{\frac{1}{\infty}} = \sqrt[\infty]{e}$. The two adjacent areas of the undefined point are not exactly equal to each other.

Inverse trigonometric functions may have undefined points.

The Neuman function is undefined at $x = 0$ when the independent variable is the x axis. The Bessel solution (function) is in the numerator of the Neuman function: $(\frac{J_\lambda \cos\lambda\pi - J_{-\lambda}}{\sin\lambda\pi}$ where J_λ is the Bessel solution when the eigenvalue is λ, and $J_{-\lambda}$ is the Bessel solution when the eigenvalue is $-\lambda$). The Neuman function may involve a principal value(s). The Hankel function is undefined at $t = 0$. The Hankel function is the sum of the Bessel solution (function) and the Neuman function that has the same Bessel solution in the numerator. The Bessel solution replaces the x axis ($x = Bessel\ solution$), and the Neuman function replaces the imaginary or y axis ($y = Neuman\ function$). The undefined point is $t = 0$ when t is the independent variable of the Bessel solution and the Neuman function. The Hankel function may involve a principal value(s). Convolutions may be approximations.

The following is an explanation of infinite series and infinite products:

An infinite series is the sum of an infinite number of numbers or functions. The first number or function has the largest numerical value. The following numbers or functions have smaller and smaller numerical values, approaching zero.

Infinite series of numbers:
(original number= $\sum_{n=0}^{\infty} C(n) = C(0) + C(1) + C(2) + C(3) + C(4) + C(5) + C(6) + \cdots\cdots$) These have an index integer(s) that increments and there is not an independent variable.

Infinite series of functions:
(original function= $\sum_{n=0}^{\infty} f(n,x) = f(0,x) + f(1,x) + f(2,x) + f(3,x) + f(4,x) + f(5,x) + f(6,x) + \cdots\cdots$)
These have an index integer(s) that increments and there is also an independent variable.

When there are fewer terms in a (infinite) series, the amount of error increases.

Despite the small amount of error, numbers, functions, or solutions of differential equations that are in infinite series form, or that can be put into an infinite series form have advantageous characteristics.

Infinite product of numbers:
(original number= $\prod_{n=0}^{\infty} C(n) = C(0)C(1)C(2)C(3)C(4)C(5)C(6)\cdots\cdots$) These have an

index integer(s) that increments and there is not an independent variable.

Infinite product of functions:
(original function= $\prod_{n=0}^{\infty} f(n,x) =$
$f(0,x)f(1,x)f(2,x)f(3,x)f(4,x)f(5,x)f(6,x)\cdots$)
These have an index integer(s) that increments and there is also an independent variable.

When there are fewer terms in an (infinite) product, the amount of error increases.

Not all numbers can be converted into infinite series form.

Not all functions can be converted into infinite series form.

Not all numbers can be converted into infinite product form.

Not all functions can be converted into infinite product form.

Some numbers in infinite series form are approximations.

Some functions in infinite series form are approximations. For example, the infinite series solutions of differential equations that are based on singularity points are not exactly equal to the solutions obtained with generating functions and/or Rodrigues' formula. Also, Taylor series are not exactly equal to the original functions.

Stirling series is an infinite series that is approximately equal to the natural logarithm of the Gamma function.

Some functions in infinite series form are not exact (disregarding the error of irrational numbers) from $-\infty$ to $+\infty$.

The following is an example of how the error of functions near boundary points may be described:
If the range of the independent variable is $-2 \le x \le +2$, then the function may be exact from -2 to $+2$,

and when the range of the independent variable is $-2 < x < +2$, then the function may (but not necessarily) have more error as the independent variable approaches -2 or $+2$.
Following is another example of how approximations may be described:

$a > C$

$a < C$

The function may (but not necessarily) have more error as a approaches C.

Iterative methods may not be exact.

Successive iterations use the last result or output for the next iteration. With each additional iteration, the result or output gets closer to the exact solution.

Newton Raphson iterations for calculating the zero points where the function crosses the horizontal axis may not be exact. Noniterative methods may be exact.

Picard successive iterations for solving ordinary differential equations may not be exact. Noniterative methods may be exact.

Application of Runge-Kutta iterations for solving differential equations may not be exact.

Many ordinary differential equations have similarities to the Sturm-Liouville equation. The following is the Sturm-Liouville equation:

$\frac{d}{dx}[c_1 y'] + c_2 y + \lambda y = c_1 y'' + c_1' y' + c_2 y + \lambda y = 0.$

Here is the rearrangement of the Sturm-Liouville equation into a well-known form:

$c_1 y'' + c_1' y' + c_2 y + \lambda y = 0$
$c_1 y'' + c_1' y' + (c_2 + \lambda) y = 0$
$c_1 y'' + c_1' y' + (c_2 + \lambda) y - (c_2 + \lambda) y = -(c_2 + \lambda) y$
$c_1 y'' + c_1' y' = -(c_2 + \lambda) y$

$$\frac{1}{-y}(c_1 y'' + c_1' y') = (-(c_2 + \lambda)y)\frac{1}{-y}$$

$$\frac{c_1 y'' + c_1' y'}{-y} = c_2 + \lambda$$

y is the solution of the differential equation. y is a function of x.

y' is the solution that is differentiated once with respect to x.

y'' is the solution that is differentiated twice (not to obtain a differential form) with respect to x.

When the solution is not an infinite series, and the independent variable x goes from $-\infty$ to $+\infty$, $c_2 + \lambda$ remains the same.

c_1, c_1', and c_2 are coefficients. Coefficients are functions or numbers that are multiplied with the solution, or the derivatives of the solution. c_1' is c_1 differentiated once with respect to x.

λ is an eigenvalue constant. The eigenvalue is also the solution index. The eigenvalues of differential equations are not the same as the eigenvalues of matrices.

Ordinary differential equations that meet Sturm-Liouville requirements can be solved with infinite series solutions, or with noninfinite series solutions obtained with generating functions and/or Rodrigues' formula.

If obtaining infinite series solutions, or noninfinite series solutions with generating functions and/or Rodrigues' formula seem complicated, solutions are available in tables.

Some ordinary differential equations may be solved with infinite series solutions near singularity points: The singularity point(s) of differential equations is the value of x that causes the coefficient of the first term to be equal to zero. For example, if the first term is $x^2 y''$, then the singularity point is $0 (0^2 = 0)$. If the first term is $(1 - x^2)y''$, then the singularity point is $\pm 1 [(1 - (\pm 1^2)) = (1 - 1) = 0]$.

If the coefficient of y' is also zero at the singularity point, then the differential equation has a regular singularity point. For example, if the singularity point is 0, and the second term is xy', then the singularity point is a regular singularity point ($xy' = 0y' = 0$).

If c_1' is not zero at the singularity point, then the differential equation has an irregular singularity point. For example, if the singularity point is ± 1, and the second term is $-2xy'$, then the singularity point is an irregular singularity point($-2(\pm 1) \neq 0$).

If the singularity point is at $x = 0$, and it is a regular singularity point and the eigenvalue is positive, then the infinite series solution will be exact at $x = 0$. As x moves away(asymptotic) from zero, the amount of error increases. If this equation meets Sturm-Liouville requirements, and λ(see Bessel equation) is a positive integer(s) from zero to infinity, then it can be solved with a noninfinite series solution obtained with a generating function. Each different eigenvalue can be paired with a different solution, and the noninfinite series solutions are exact from $-\infty$ to $+\infty$.

If the singularity point is ± 1, and it is an irregular singularity point and the eigenvalue is a positive integer(s), then the infinite series solution will be undefined at ± 1, but the infinite series solution will be exact at $x = 0$. If this equation meets Sturm-Liouville requirements, and λ(see the Legendre equation) is a positive integer(s) from zero to infinity, then it can be solved with a noninfinite series solution obtained with a generating function and/or Rodrigues' formula. Each different eigenvalue can be paired with a different solution. The noninfinite series solutions are exact at ± 1, and from $-\infty$ to $+\infty$.

It is impossible to exactly symbolically solve the associated Legendre equation.

Infinite series solutions that are based on singularity points of differential equations are not exactly the same as the noninfinite series solutions obtained with generating functions and/or Rodrigues' formula.

Not all differential equations can be solved with infinite series solutions. It may not be possible to solve, or exactly solve some differential equations. Some differential equations can be exactly solved with noninfinite series solutions obtained with generating functions and/or Rodrigues' formula.

It is not possible to obtain generating functions for all differential equations.

It may not be possible to use generating functions and/or Rodrigues' formula to solve (or exactly solve) differential equations when λ is not a positive integer(s).

Differential equations may need to be multiplied with weight functions to be solvable. Weight functions are functions with the same independent variable. The weight function is multiplied with each term of the differential equation so that the differential equation can be solved with an infinite series solution, or with a noninfinite series solution obtained with a generating function and/or Rodrigues' formula.

Differential equations that are multiplied with weight functions are not exactly the same as the original differential equations.

Finite difference method for differential equations and partial differential equations may not be exact.

Numerical calculations for derivatives may not be exact. Multivariate functions can be numerically differentiated with the finite difference method. The functions to be differentiated are usually already known. Multiple numerical differentiations can also be accomplished with the finite difference method. Following is $y(x)$ numerically differentiated with the finite difference method:

$\frac{dy(x)}{dx} = \frac{y(x+\Delta x)-y(x)}{\Delta x}$ where Δx(pronounced delta x) is a small difference in x

Some differential equations are chaotic.

Bessel equation(ordinary differential equation) is: $x^2 y'' + xy' + (x^2 - \lambda^2)y = x^2 y'' + xy' + x^2 y - \lambda^2 y = 0$, where $c_1 = x^2$, the coefficient of y' is x, other terms of the Bessel equation and the Bessel solutions compensate for the coefficient of y' because it is not equal to the derivative of $c_1 (c_1' = \frac{d}{dx} x^2 = 2x \neq x =$ coefficient of y'), and $c_2 = x^2$.

Legendre polynomials. These are exact noninfinite series solutions obtained with a generating function and/or Rodrigues' formula. The Legendre ordinary differential equation is: $(1 - x^2)y'' - 2xy' + \lambda(\lambda + 1)y = (1 - x^2)y'' - 2xy' + (\lambda^2 + \lambda)y = (1 - x^2)y'' - 2xy' + \lambda^2 y + \lambda y = 0$, where $c_1 = 1 - x^2$, $c_1' = -2x$, and $c_2 = \lambda^2$. The Legendre differential equation can also be solved with infinite series solutions.

Airy equation(ordinary differential equation) is: $y'' - xy = 0$, where $c_1 = 1, c_1' = 0, c_2 = x$. The Airy equation has only one eigenvalue($\lambda = 0$).

Chebyshev polynomials. These are exact noninfinite series solutions obtained with a generating function and/or Rodrigues' formula. The Chebyshev ordinary differential equation is: $(1 - x^2)y'' - xy' + \lambda y = 0$ where $c_1 = 1 - x^2$, the coefficient of y' is $-x$, the other terms of the Chebyshev equation and the Chebyshev solutions compensate for the coefficient of y' because it is not equal to the derivative of $c_1 (c_1' = \frac{d}{dx}(1-x^2) = -2x \neq -x = coefficient\ of\ y')$, and $c_2 = 0$. The Chebyshev differential equation can also be solved with infinite series solutions.

Laguerre polynomials. These are noninfinite series solutions obtained with a generating function and/or Rodrigues' formula. The Laguerre ordinary differential equation $(xy'' + (1-x)y' + \lambda y = 0)$ need to be multiplied with a weight function because the coefficient of y' is not equal to the derivative of $c_1 (c_1' = \frac{d}{dx} x = 1 \neq 1 - x = coefficient\ of\ y')$. The Laguerre ordinary differential equation can also be solved with infinite series solutions.

Hermite polynomials. These are noninfinite series solutions obtained with a generating function and/or Rodrigues' formula. The Hermite ordinary differential equation $(y'' - 2xy' + 2\lambda y = 0)$ need to be multiplied with a weight function because the coefficient of y' is not equal to the derivative of $c_1 (c_1' = \frac{d}{dx} 1 = 0 \neq -2x = coefficient\ of\ y')$. The Hermite ordinary differential equation can also be solved with infinite series solutions.

Gauss error correction function cannot be exactly symbolically calculated.

Euler-Maclaurin integrations may not be exact.

$(\frac{\delta}{\delta x})$ means to partially differentiate the following multivariate function with respect to x.

Not all partial differential equations can be solved, or exactly solved.

Solutions for partial differential equations that are obtained with the chain rule $(\frac{df(g(t))}{d(g(t))} \frac{d(g(t))}{dt} = \frac{df(g(t))}{dt}$

change of variable and $g(t)$ is a vector) may not be exactly the same as solutions obtained with respect to the original variables ($\frac{df(x)}{dx}$).

The following is the Laplace partial differential equation:

$\frac{\delta^2 f(x,y,z)}{\delta x^2} + \frac{\delta^2 f(x,y,z)}{\delta y^2} + \frac{\delta^2 f(x,y,z)}{\delta z^2} = 0$ where $f(x, y, z)$ is the solution. $f(x, y, z)$ is partially differentiated twice(not to obtain a differential form) with respect to x, partially differentiated twice(not to obtain a differential form) with respect to y, and partially differentiated twice with respect to z.

The Laplace partial differential equation can be solved with the separation of variables method. Separation of variables method converts partial differential equations into ordinary differential equations. The separation of variables method is exact. Not all partial differential equations can be solved with the separation of variables method.

Finite element method for solving partial differential equations may not be exact. In the finite element method, the (exact) solution is usually already known.

Functions that are made orthogonal are not exactly the same as the original functions.

Normalized functions are not exactly the same as the original functions.

Calculus of residues involving undefined points and limits may not be exact.

The following is the equation of a circle with a radius of 5 and its rearrangement for differentiation and also for integration:

$y^2 + x^2 = r^2$
$y^2 + x^2 = 5^2$
$y^2 + x^2 = 25$

$$y^2 = 25 - x^2$$
$$\sqrt{y^2} = \sqrt{(25 - x^2)}$$
$$y = \sqrt{(25 - x^2)}$$
$$\frac{d}{dx}y = \frac{d}{dx}\sqrt{(25 - x^2)}$$ for differentiation(similarly,
$$\frac{d}{dy}x = \frac{d}{dy}\sqrt{(25 - y^2)})$$
$\int y\,dx = \int\sqrt{(25 - x^2)}dx$ for integrating(similarly, $\int x\,dy = \int\sqrt{(25 - y^2)}dy$) Try those on a symbolic internet calculator.

If symbolically obtaining $\frac{d}{dx}\sqrt{(25 - x^2)}$ or $\frac{d}{dy}\sqrt{(25 - y^2)}$ is not possible, it may be possible to differentiate with the chain rule. Derivatives obtained with the chain rule($\frac{df(g(t))}{d(g(t))}\frac{d(g(t))}{dt} = \frac{df(g(t))}{dt}$ change of variable and $g(t)$ is a vector) may not be exactly the same as derivatives obtained with respect to the original variables ($\frac{df(x)}{dx}$).

If symbolically obtaining $\int\sqrt{(25 - x^2)}dx$ or $\int\sqrt{(25 - y^2)}dy$ is not possible, it may be possible to integrate by substitution. The original independent variable is replaced with another function. Integrations by substitution (change of variable) may not be exactly the same as integrations with respect to the original variables.

Circles, spheres, cylinders, or cylindrical objects (radius is not constant) may need to be differentiated with the chain rule, or integrated by substitution.

Following is the equation of an ellipse that is centered at the origin. The distance from the center to the left side or the right side is 5, and the distance from the center to the top or the bottom is 4:

$$\frac{x^2}{5^2} + \frac{y^2}{4^2} = 1$$

$$\frac{x^2}{25} + \frac{y^2}{16} = 1$$

The following is the rearrangement of the above ellipse equation for integration:

$$\frac{x^2}{25} + \frac{y^2}{16} = 1$$

$$\frac{y^2}{16} = 1 - \frac{x^2}{25}$$

$$y^2 = 16(1 - \frac{x^2}{25})$$

$$y = \sqrt{(16(1 - \frac{x^2}{25}))}$$

$\int y\, dx = \int (\sqrt{16(1 - \frac{x^2}{25})})dx$ (similarly, $\int x\, dy =$ $\int (\sqrt{16(1 - \frac{y^2}{25})})dy)$

Try those on a symbolic internet calculator. It may be necessary to replace the independent variables with other functions to symbolically integrate elliptical functions.

Differential form of circles ($d(x^2 + y^2 = r^2) =$ $-\frac{y}{x^2+y^2}dx + \frac{x}{x^2+y^2}dy$) is closed and not exact. The differential form of circles can be analyzed with cohomological methods.

When differentiating $x^2 + y^2 = r^2$ with respect to x, the differential form is approximately $-\frac{y}{x^2+y^2}dx$, and when differentiating $x^2 + y^2 = r^2$ with respect to y, the differential form is approximately $\frac{x}{x^2+y^2}dy$.

A possible contravariant tensor (vectors) of circles is: $-\frac{y}{x^2+y^2}dx + \frac{x}{x^2+y^2}dy$.

A possible covariant tensor (vectors) of circles is:

$$\begin{array}{c} -\frac{y}{x^2+y^2}dx \\ \frac{x}{x^2+y^2}dy \end{array}.$$

The Maxwell mnemonic B is a three-dimensional electromagnetic field, and the Maxwell mnemonic E is a three-dimensional electrostatic field.

A possible contravariant tensor (vectors) of electromagnetic fields is: $\frac{d}{dx}B + \frac{d}{dy}B + \frac{d}{dz}B$.

A possible covariant tensor (vectors) of electromagnetic fields is: $\begin{array}{c}\frac{d}{dx}B \\ \frac{d}{dy}B \\ \frac{d}{dz}B\end{array}$.

For less missing information, for example, the vector $\frac{d}{dx}B$ at $y = 0$ can be replaced with the vector $\frac{d}{dx}B$ at $y = -6, -5, -4, -3, -2, -1, 0, 1, 2, 3, 4, 5, 6$. For even less missing information, for example, the vector $\frac{d}{dx}B$ at $y = 0$ can be replaced with the vector $\frac{d}{dx}B$ at $y =$
$-6, -5\frac{1}{2}, -5, -4\frac{1}{2}, -4, -3\frac{1}{2}, -3, -2\frac{1}{2}, -2, -1\frac{1}{2}, -1, -\frac{1}{2}$,
$0, \frac{1}{2}, 1, 1\frac{1}{2}, 2, 2\frac{1}{2}, 3, 3\frac{1}{2}, 4, 4\frac{1}{2}, 5, 5\frac{1}{2}, 6$, and similarly for $\frac{d}{dy}B$ and $\frac{d}{dz}B$.

A possible contravariant tensor (vectors) of electrostatic fields is: $\frac{d}{dx}E + \frac{d}{dy}E + \frac{d}{dz}E$.

And a possible covariant tensor (vectors) of electrostatic fields is: $\begin{array}{c}\frac{d}{dx}E \\ \frac{d}{dy}E \\ \frac{d}{dz}E\end{array}$.

An Introduction to Manufacturing and Mathematical Limitations

For less missing information, for example, the vector $\frac{d}{dx}E$ at $y = 0$ can be replaced with the vector $\frac{d}{dx}E$ at $y = -6, -5, -4, -3, -2, -1, 0, 1, 2, 3, 4, 5, 6$. For even less missing information, for example, the vector $\frac{d}{dx}E$ at $y = 0$ can be replaced with the vector $\frac{d}{dx}E$ at $y =$
$-6, -5\frac{1}{2}, -5, -4\frac{1}{2}, -4, -3\frac{1}{2}, -3, -2\frac{1}{2}, -2, -1\frac{1}{2}, -1, -\frac{1}{2},$
$0, \frac{1}{2}, 1, 1\frac{1}{2}, 2, 2\frac{1}{2}, 3, 3\frac{1}{2}, 4, 4\frac{1}{2}, 5, 5\frac{1}{2}, 6$, and similarly for $\frac{d}{dy}E$ and $\frac{d}{dz}E$.

If symbolically obtaining $\frac{df(x)}{dx}$ is not possible, it may be possible to differentiate with the chain rule. Derivatives obtained with the chain rule($\frac{df(g(t))}{d(g(t))}\frac{d(g(t))}{dt} = \frac{df(g(t))}{dt}$ change of variable and $g(t)$ is a vector), may not be exactly the same as derivatives obtained with respect to the original variables ($\frac{df(x)}{dx}$).

If symbolically obtaining $\int f(x)\,dx$ is not possible, it may be possible to integrate by substitution. The original independent variable is replaced with another function. Integrations by substitution (change of variable) may not be exactly the same as integrations with respect to the original variables. The following is an example of integration by substitution: $\int f(g(t))dg(t)$ where $g(t)$ is a vector, and in symbolic integration, $dg(t)$ means to differentiate the function $g(t)$.

Functions that are modified with another function (integrating factor) so they can be symbolically integrated are not exactly the same as the original functions.

Fourier series transformations convert functions into series terms. The first term has a cosine function in it. The following terms have cosine functions or sine functions in them.

Not all functions can be Fourier series transformed. When there are fewer terms in a Fourier series, the amount of error increases.

Gibbs phenomenon is the error of Fourier series transformation of square waves.

Fourier series involve an averaging method. For example:

$Fourier\ coefficient\ for\ the\ nth\ term\ of\ the\ series =$
$\frac{1}{D}\int_0^D f(t)\cos(\frac{n\pi t}{D})dt$ where in symbolic integration, dt means to differentiate the function t.

$Fourier\ coefficient\ for\ the\ nth\ term\ of\ the\ series$ is the area between $f(t)\cos(\frac{n\pi t}{D})$ and the horizontal axis from 0 to D, divided by D.

$Fourier\ coefficient\ for\ the\ nth\ term\ of\ the\ series$ is a length or distance.

The following are some other examples of averaging methods:

$\frac{(x+y)}{2}$

$\sqrt{(xy)}$

$\frac{1}{\pi}\int_0^\pi f(t)dt$

$\frac{1}{2\pi}\int_{-\pi}^{+\pi} f(t)dt$

Obtaining the tangent at the midpoint of a segment of a curved function.

Obtaining the normal at the midpoint of a segment of a curved function.

Obtaining the integral at the midpoint of a segment of a curved function.

Obtaining the derivative at the midpoint of a segment of a curved function.

One of the disadvantages of averaging methods is that different input values may produce the same average value. For example, $\frac{5+5}{2} = 5$, and $\frac{6+4}{2} = 5$. Also $\sqrt{(5)(5)} = 5$(arithmetic mean), and $\sqrt{(12.5)(2)} = 5$.

Schläfli integral involves an averaging method. The following is an example of the Fourier integral transform equation (not infinite series): $FT(s) = \frac{1}{2\pi \text{ (or } \sqrt{2\pi})} \int_{-\infty}^{\infty} f(t)e^{0-ist} dt$ where in symbolic integration, dt means to differentiate the function t. The $d\_$ notation also indicates which variable varies from $-\infty$ to $+\infty$. e^{-it} is equal to $cosine\ t - isine\ t$, where the function $cosine\ t$ replaces the x axis ($x = cosine\ t$), and the function $isine\ t$ replaces the y or z axis ($y = sine\ t$, or $z = sine\ t$). These functions draw a circle that is centered at the origin ($x = 0, y = 0$), with a radius of one. $FT(s)$ is a function of s. It usually does not contain the variable t. $FT(s)$ is equal to the area between $f(t)e^{0-ist}$ and the horizontal axis from $-\infty$ to $+\infty$, and divides the area by 2π or $(\sqrt{2\pi})$. $FT(s)$ is a length or distance (or it is related to length or distance).

The Fourier integral transformation equation (not infinite series), have two variables. The variable t goes from negative infinity to positive infinity. The variable s is constant during integration. There may be a third variable that is just a constant. Some Fourier integral transformations are approximations. When the variable s moves away from the undefined point, the amount of error will decrease, and as s goes to infinity, the amount of error may not decrease to absolute zero except at $s = -\infty$ and $s = +\infty$.

Not all functions converge enough to be Fourier integral transformed. An example of a function that converges is $\frac{1}{x}$. As x approaches $\pm\infty$, the function $\frac{1}{x}$ approaches zero. The function $\frac{1}{x^2}$ will converge more than the function $\frac{1}{x}$.

Not all functions that converge enough can be symbolically Fourier integral transformed.

The following is an example of the Laplace integral transform equation: $LT(s) = \int_0^\infty f(t)e^{-st+i0}dt$ where in symbolic integration, dt means to differentiate the function t. The $d\_$ notation also indicates which variable varies from 0 to ∞. $LT(s)$ is a function of s. It usually does not contain the variable t. $LT(s)$ calculates the area between $f(t)e^{-st+i0}$ and the horizontal axis from 0 to ∞.

The Laplace integral transformation equation have two variables. The variable t goes from zero to infinity. The variable s is constant during integration. There may be a third variable that is just a constant. Some Laplace integral transformations are approximations. When the variable s moves away from the undefined point, the amount of error will decrease, and as s goes to infinity, the amount of error may not decrease to absolute zero except at $s = -\infty$ and $s = +\infty$. For example, when $LT(s) = \int_0^\infty (\sinh t)e^{-st+i0}dt = \frac{c}{s^2-c^2}$, the undefined points are $s = \pm C$.

The following is the Laplace inverse integral transform equation: $f(t) = \int_0^\infty LT(s)e^{-st+i0}ds$ where in symbolic integration, ds means to differentiate the function s. In the Laplace inverse integral transform equation, the undefined points are in the function that is integrated.

Not all functions converge enough to be Laplace integral transformed.
Not all functions that converge enough can be symbolically Laplace integral transformed.
Application of Mellin integral transforms may not be exact.
The following is the Taylor series equation: $f(x) = \sum_0^\infty \frac{d^n f(x)}{n!}(x-c)^n$, where c is the center (not necessarily at $x = 0$), and n is also the number of times the function $f(x)$ is differentiated with respect to the independent variable x. For example:

$f(x) = \sum_0^\infty \frac{d^n f(x)}{n!}(x-c)^n = f(x) + \frac{d^1 f(x)}{1!}(x-c)^1 + \frac{d^2 f(x)}{2!}(x-c)^2 + \frac{d^3 f(x)}{3!}(x-c)^3 + \frac{d^4 f(x)}{4!}(x-c)^4 + \frac{d^5 f(x)}{5!}(x-c)^5 + \frac{d^6 f(x)}{6!}(x-c)^6 + \cdots\cdots$ where the first term is $f(x)$ because $0! = 1$, and $(x-c)^0 = 1$

The center point of Taylor series, (infinite or not infinite, with nonzero or zero remainders) is exact. When the independent variable moves away from the center point of a Taylor series (infinite or not infinite, with nonzero or zero remainders), the amount of error increases.

When $x = c$, then $(x-c)^n = (c-c)^n = 0^n = 0$, and the above Taylor series becomes:

$f(c) = f(c) + 0 + 0 + 0 + 0 + 0 + \cdots\cdots$

It is not possible to convert functions that do not converge enough into Taylor infinite series, but it may be possible to convert functions that do not converge enough into Taylor series (not infinite with remainders of zero).

When there are fewer terms in a Taylor (infinite) series, the amount of error increases.

Maclaurin series is the same as Taylor series except the center of the series is at $x = 0$.

It may be possible to square root negative numbers if circumstances allow the use of imaginary numbers. For example, $(i2)(i2) = (i2)(-i2) = (-i2)(i2) = (-i2)(-i2) = -4$.

The equation for solving quadratic equations ($ax^2 + bx + c = 0$) does not work when the value that is underneath the square root sign is negative ($x = \frac{-b \pm \sqrt{b^2 - 4ac}}{2a}$).

The following is an example of a quadratic equation that is formed with the FOIL method(product of (F)irst terms plus the product of (O)uter terms plus the product of (I)nner terms plus the product of (L)ast terms):

$(x + 1)(x + 10) - (6)(6) = 0$

$((x)(x) + (10)(x) + (1)(x) + (1)(10)) - 36 = 0$

$(x^2 + 10x + 1x + 10) - 36 = 0$

$(x^2 + 11x - 26) = 0$ where $a = 1, b = 11$, and $c = -26$

The solutions are $x = \frac{-b \pm \sqrt{b^2 - 4ac}}{2a} = \frac{-11 \pm \sqrt{(11^2) - (4)(1)(-26)}}{(2)(1)} = \frac{-11 \pm \sqrt{(121 + 104)}}{2} = \frac{-11 \pm \sqrt{225}}{2} = \frac{-11 \pm 15}{2} = \frac{4}{2}$ or $\frac{-26}{2} = 2$ or -13

The zeros(solution) of the function are $x = 2$ and $x = -13$. The zeros of a function can also be the eigenvalues of a matrix. See matrix eigenvalue above.

Matrix eigenvalues that are the zeros of a function obtained with the Newton Raphson successive iteration method may not be exact.

Matrix eigenvalues obtained with the Gershgorin method may not be exact.

Not all functions can be symbolically square rooted. For example, the square root of x^2 is x, but it may not be possible to symbolically square root $25 - x^2$ exactly.

Not all functions can be symbolically cube rooted. For example, the cube root of x^3 is x, but it may not be possible to symbolically cube root $25 - x^2$ exactly.

It is not possible to obtain the logarithm (Log) of negative numbers. The logarithm of an original number is the exponent value of 10^x, where x is a real number(not imaginary) that makes 10^x equal to the original number. For example,
$Log(original\ number) = Log(10) = 1$,
$Log(original\ number) = Log(100) = 2$,
$Log(1000) = 3$, and $Log(10,000) = 4$. Since 10^x is always positive, it is not possible to obtain the logarithm of negative numbers.

Also, it is not possible to obtain the natural logarithm (Ln) of negative numbers. The natural logarithm of an original number is the exponent value of e^x, where x is a real number(not imaginary) that makes e^x equal to the original number. For example, $Ln(original\ number) = Ln(e^1) = 1$, $Ln(original\ number) = Ln(e^2) = 2$, $Ln(e^3) = 3$, and $Ln(e^4) = 4$. Since e^x is always positive, it is not possible to obtain the natural logarithm of negative numbers.

Interpolations may not be exact.
Polynomial interpolations may not be exact.
Application of the minimum principle may not be exact.
Application of the minimax theorem may not be exact.
Strong approximation theorem.
Monte Carlo method may not be exact.
Application of meromorphic functions that may need Cauchy principal values may not be exact.
Application of the L-function may not be exact.

Some applications of functions that have an inner product may not be exact.
Functions that are normed may not be exactly equal to the original functions.
The absolute value function (| |) removes the negative sign on numbers.
Following is an example of a normed function:
$$\sqrt{x^2 + |(iy)^2|} = \sqrt{x^2 + y^2} =$$
$normed\ function\ coordinate\ value$
x is the real part of a function coordinate, and iy is the imaginary part of a function coordinate. An infinite number of possible x and iy coordinate values will produce the same
$normed\ function\ coordinate\ value$ (infinite dimensions).
Some applications of the Hilbert space method may not be exact.
Following is a simplified example of the Hilbert space method, where an original infinite series is converted into an approximate Hilbert space equivalent:
$(original\ infinite\ series) = \sum_{n=0}^{\infty} f(n,x) = f(0,x) + f(1,x) + f(2,x) + f(3,x) + f(4,x) + f(5,x) + f(6,x) + \cdots) \cong (Hilbert\ space\ equivalent) = (g_0(x))(g_0(x)) + (g_1(x))(g_1(x)) + (g_2(x))(g_2(x)) + (g_3(x))(g_3(x)) + (g_4(x))(g_4(x)) + (g_5(x))(g_5(x)) + (g_6(x))(g_6(x)) + \cdots$
Not all functions can be exactly symbolically square rooted. It may be necessary to replace the variables with numbers and numerically square root.
Not all squared functions can be exactly symbolically integrated.
Least squares method may not be exact. The following is a simplified example of the least squares method:

$\int_a^b (f(x) - g(x))^2 dx = output$, where $f(x)$ is exact and $g(x)$ is an approximation. The goal is to choose $g(x)$ for the smallest(least) output.

Euler's method of using circles to simulate curved segments of functions may not be exact.

Born approximation

Mathematics involving variations may not be exact.

Application of the WKB method for solving differential equations may not be exact.

Steepest descent method may not be exact.

Stationary phase method may not be exact.

Weierstrass approximation

Frenet equation may not be exact when numerically calculated.

Padé method may not be exact.

Rayleigh-Ritz method may not be exact.

Application of the Sylvester equation may not be exact. The following is a simplified explanation of the Sylvester equation:

$(product\ of\ the\ solution\ and\ the\ row\ terms\ of\ matrix\ A) +$
$(product\ of\ the\ same\ solution\ and\ the\ column\ terms\ of\ mat$
$output\ matrix$

Application of the Lyapunov equation may not be exact.

Krylov method may not be exact.

Jacobi iteration may not be exact.

Hooke's law may not be exact.

Application of the Galerkin method may not be exact.

Application of spectral sequences may not be exact. Spectral sequences are successive iterations.

Application of simplicial complexes may not be exact.

Curved surfaces formed with many small triangles may not be exact.

Application of CW complexes may not be exact.
Curved surfaces formed with many small circles may not be exact.
Some symbolic entries in Fourier series transformation tables may be approximations.
Some symbolic entries in Fourier integral transformation tables may be approximations.
Some symbolic entries in Laplace integral transformation tables may be approximations.
If there is not an algebraic method to get from one side of an equal sign to the other side of the equal sign, the equation or function may be an approximation.
If an equation is modified to obtain a solution, the solution may not be exact with respect to the original equation.
The sun is not perfectly spherical.
The Earth and the other planets that revolve around the sun are not perfectly spherical.
The orbits of the Earth and the orbits of the other planets that revolve around the sun are not perfect circles or perfect ellipses.
The moon is not perfectly spherical.
The orbit of the moon around the Earth is not a perfect circle or a perfect ellipse.
If the mathematics is not exact, slight adjustments can be made after calculations, or on the mechanical or electrical products.

Recommended reading:

This book is not comprehensive. Read on the internet and in other books for a better understanding of manufacturing and mathematical limitations.

www.ingramcontent.com/pod-product-compliance
Lightning Source LLC
Chambersburg PA
CBHW031538210526
45464CB00003B/1059